more things in heaven and earth

For Allie-

Friend, poet, on her way -

Kurt

ALSO BY KURT BROWN

Poetry
Return of the Prodigals

Chapbooks
The Lance & Rita Poems
Recension of the Biblical Watchdog
A Voice in the Garden: Poems of Sandor Tádjèck
Mammal News

Editor
The True Subject
Writing It Down for James
Facing the Lion
Drive, They Said: Poems about Americans and Their Cars
Night Out: Poems about Hotels, Motels, Restaurants and Bars
Verse & Universe: Poems about Science & Mathematics
The Measured Word: On Poetry & Science

more things in heaven and earth

kurt brown

four way books

new york, new york

Editorial Office
Four Way Books
PO Box 535, Village Station
New York, NY 10014
www.fourwaybooks.com

Library of Congress
Catalog Card Number: 00 134375

ISBN 1-884800-34-3

Book Design: Brunel
Cover Art: Frits van den Berghe, *L'homme des nuages,*
from the collection of the Musée Royeaux des Beaux-Arts de Belgique, Bruxelles

This book is manufactured in the United States of America and printed on acid-free paper.

Four Way Books is a division of Friends of Writers, Inc., a Vermont-based not-for-profit organization. We are grateful for the assistance we receive from individual donors and private foundations.

ACKNOWLEDGMENTS

My thanks to the editors of the following reviews, in which these poems appeared, some in earlier versions:

Agni	A New Age
	Banana Knife
The Atlanta Review	White, Middle-Class, Male
Bottomfish	Peter Lorre on the Porch
	Massive
	Aliens
Connecticut Review	Fisherman
Crab Orchard Review	Wow & Flutter
Crazyhorse	Right & Wrong
	A Father's Joke
	The American Helmet in Provence
The Harvard Review	At the retirement home for slang
High Plains Literary Review	How It Arrives
Luna	Master of Consciousness
	Serial Position Effect
The Marlboro Review	The Good Devil
Pavement Saw	Dear Reader, I Mean No Harm
¶ A Magazine of Paragraphs	One of the Spared
Ploughshares	In the Library of Unrecorded Lives
	A '49 Merc
Rattle	At the Health Club
Salamander	Census
The Tampa Review	Reflections of Notre Dame de Paris
Tar River Poetry	Spreading the Word
Virginia Quarterly Review	The Committee to Upgrade Celestial Signs

"The Good Devil," appeared in *Outsiders: Poems about Rebels, Exiles and Renegades*, edited by Laure-Anne Bosselaar, Milkweed Editions, 1999.

SPECIAL THANKS

Tony Hoagland: commentary in time
Thom Ward: vital energy
Tom Lux: an uncompromising eye
Stephen Dobyns: continuing support

My wife, Laure-Anne, from beginning to end

An extra-special thanks to Martha Rhodes and Jean Brunel

CONTENTS

In the Library of Unrecorded Lives

Dear Reader, I Mean No Harm

More Things in Heaven and Earth

In the Library of Unrecorded Lives

In the Library of Unrecorded Lives

If there's one good book in everyone, as the saying goes,
at least one, I like to think of Miss Shaughnessy,

my third-grade teacher, hair bunched back in a crimped bun,
dress as drab as those gray afternoons spent dreaming

in her class, eyes two magnified minnows behind glass
thick as astronomical lenses—could she have had a hidden life?

Could she have dreamed of swashbucklers and maids,
murder and sex merging in a narrative so gripping

little boys would stop griping and fall silent under its spell?
And what of Rosa, Rosa Puckett, face pale as oatmeal,

soft sac of silence we attacked like hornets, beleagured figure
of fourth-grade fun? Could she have harbored romance,

longed behind that shabby scowl for some exotic rendezvous
on foreign sand? Or Tony Conecci, block of meat,

ham-fisted son of immigrants—could he have imagined
heroes with a swagger, cloak-and-dagger exploits

beyond the weekend pounding he withstood as center
on the football squad? What of their lives and so many others

God seemed loath to honor, nameless lumpen
in the pageant of my own pubescent brain? Yet here they are,

radiant in memory, each a star in the mind's
incorruptible eye: Nero Legér, flashing horse teeth

when he laughed, arms jutting out of tattered cuffs;
Peggy Hood, so shy her feet scuffed circles when she spoke,

butt of every joke we could muster, which only made her stammer
worse. What of their lives, unrecorded in any book?

I'd love to slip their volumes off the shelf and have a look,
settle in to read the wealth of life denied—learn

what Andy Washburn felt the night his ailing mother died,
who coughed up fifty years of smoke and vanished;

or Cindy Fitz, banished on a withered leg, tall as a fencepost
and as lonely. Ordinary figures of disappointment and strife.

If there's one good book in everyone, I'd love to browse
those pages after years of currying my own regard.

Lucent populations of the soul: it's for you I set down my life.

Family Gods

Rifling my parents' clothes for cash,
I came upon it one rainy morning
in the warrens of their closet
stuffed deep in a pocket of my father's coat.
Sex was a secret in our house, nothing
to speak of in those lurid summers—
sweat made your shirt stick
and roads melted like spent tallow.
Father was away, a merchant sailor
bound for explicit ports of call.
Mother worked all day selling burgers
and antiques, sometimes a bottle
of cheap rum in the town's only emporium.
Now here I was, furtive thief
trying one pocket, then another,
until I plucked it out: more-than-life-sized
and obscenely veined. It was molded
of firm white rubber, a perfect phallus
in all detail—turgid, arched,
with broad shaft and hooded glans.
I held it like a person holds
a bomb—revolved it lightly in my palm
to see its other end—abrupt stump
detached from any body.
I stared until I felt it stir, as though
I'd found one of the family gods.
Then hurriedly stuffed it back
into his coat, more potent than money,
more explosive, and went on looking
for loose change. So heinous a crime,
so deeply buried in my brain
I might have made it up out of
sex-starved, adolescent fantasies—
though my hand still tingles and I'm antsy,

even now, speaking of it—
my parents, both dead,
might surprise me any minute,
punish me for announcing to the world
their secret and delicious love.

The Hots

I've got the hots for her . . .
that's what we used to say.

She's hot to trot, meaning
long, sleek legs like a thoroughbred?

I'm hot on her trail, hound
and fox—meaning fox is dainty,

fox is ladylike in lean pursuit.
I've got the hots, as if it were a pox,

a carnal plague that raised our blood
and flushed our cheeks.

The flames that we imagined
mirrored hell, that place

the preacher advertised each week
to cool our lust. *She's hot stuff*

like lava or uranium,
anything explosive and about to blow.

Though what blew up was us,
teenage guages topping out,

rockets on a launch pad
sheathed in smoke *all systems go*

A Father's Joke

Hand in hand he took them to the subway,
his three lost children abandoned each year
when he sailed off, Captain of a merchant ship,
hurried them into the harrowing earth

below Manhattan where darkness echoed
and trains slithered up like serpents
to gobble hordes of passengers then rattle off
scattering clouds of scorched air.

Inside the car, they'd sit beside him
swaying as the train lurched through tunnels
lit by flashing lights and bells
like funerals for the damned.

But soon he'd turn to face them, cold-eyed,
deep in his burlesque, forehead
furrowed with a question he proposed:
"Who are you?" he'd say. *"I* don't know *you!"*

Then he'd stare as though he meant it,
leaning back to register his scorn.
All their protestations got them nowhere
as the train plunged deeper

into blackness, picking up alarming speed.
"It's us," they'd bleat, "Daddy, it's us!"
"No," he'd say. "I don't know you," and he'd
turn away, composed as any stranger.

Once he even got up and, safe on sea legs,
wavered down the aisle to take
a seat and stare serenely out the window
as though he wasn't there.

On the edge of their terror, he'd relent,
come sit by them again and grin,
arm thrown carelessly around shoulders
still shuddering with fright.

Later, in the upper world of searing light,
he'd treat them to milkshakes, still laughing,
glad to be alive: a father once again,
Captain of their dark, adoring eyes.

Peter Lorre on the Porch

Many of you don't know him now—
pale imp who slurred his words,
round eyes drooping in a beaten-puppy look
over curled, rubbery lips that sneered—

or know the song he sang, a standard then,
though singing wasn't what he did, exactly,
throttling the lyrics up through his throat
like some snot-nosed baby in its crib:

My old flame, I can't even remember her name . . .
hour after hour on my family's front porch
while summer passed unnoticed
stirring the green neighborhood of leaves.

I would slide the brittle record from its sheath,
as if it were an ancient scroll, and place it
softly on the metal wheel of my parents' Victrola,
a venerable box oracular with song,

then swing the double-jointed arm across
and set the needle in its groove—
the room filled with bland static
that drowned the rasping of locusts outside.

But soon, a bright announcement
of trumpets, swallowed up by strings;
while underneath the woodwinds pulsed—
a long, pathetic groan of saxophones.

And then that voice, *his* voice, half sob half
snarl, sniveling its moist lament:
How it drew me into its odd torture!
My old flame, I can't even remember her name . . .

And finally, the kicker—they'd engineered
a fire on the second chorus, the seethe
and crackle of flames that grew louder as he sang
until his voice rose in quickened terror:

My old flame, I can't even remember her name! . . .
Almost screaming over the rising flames,
his voice reverberating as he sank
like some damned soul into the chasms of hell.

How could I have known what love would do—
years later blubbering like a baby, eyes
burning, the taste of sulfur on my tongue,
turning on the spit of my own ridiculous pain.

Endless Sleep

That was the summer of "Endless Sleep," one of those dopey
    adolescent disaster songs,
electric bass and guitar pounding with an ominous beat,
    half arythmic, that sounded
like a funeral march *ta DUM dum dum dum / ta DUM dum dum dum*
    How my heart stepped
to that gloominess, that solemn procession of morbid notes
    which struck a chord somehow
in my teenage psyche, casting its shadow into every corner of my life.

You know the story? Some guy is downcast, desperate
    because he's argued with his girl.
In a transport of sorrow she deserts him, takes a walk one night
    along the sea, all the claptrap
of Gothic romance trundled out to help create a mood
    of uncertainty and fear:

The night was black / rain fallin' down
looked for my baby / she's nowhere around

his voice reverberant,
suggesting solitude, some hollowed-out place in the abandoned heart.

Why did we quarrel / why did we fight
why did I leave her / alone tonight?

an ecstasy of guilt
I somehow understood, though I can't recall betraying anyone,
    not then at least, still young enough
to believe in the sanctity of love, ravishment unburdened
    by its flaws, perfect love
that harbors a reproach to adults and their bitter quarrelling—
    if argument is a form of betrayal,
a breach in some ideal of absolute harmony and bliss.

Traced her footsteps / down to the shore
'fraid she's gone / forevermore

All summer those dark words
pounded my brain, though it seemed a summer more beautiful
than most, tall trees
billowing with leaves, day after day of immaculate weather.
I spent hours fishing from a rock
in the middle of the river, transistor radio perched beside me
blasting that narrative
of Eros and Thanatos over the water crowding me on every side.

I looked at the sea / and it seemed to say
I took your baby / from you away

What did it mean,
that abyss of water, a single wave reaching out to seize a woman,
drag her into depths
of cold obliteration that might wash through her to relieve the pain?
Of course he saves her,
plunges in to haul her body from the waves, and so avoids a life
of misery and regret

My heart cried out / she's mine to keep
I saved my baby / from an endless sleep.

Who knows what they were arguing about? Each day I pulled fish
out of the water like bright answers,
unable to connect that grave foreboding to the day's abundance,
its blowing light. The future
spread before me like the surface of the river, beautiful but opaque,
those lyrics black as flies
dancing in lucid air, while I had visions of a steadfast life,
a fervent life free of bitterness
and the poison of betrayal. I would never destroy what I loved.

Spreading the Word

Knock knock it's the Jesuits next door
come to take possession of the books I offer,
a complete set of early church fathers:
Justin, Gregory, Augustine, Jerome.
They file in, all nods and smiles, white shirts and ties.
Each pair of open arms receives the Word,
stacked dozens high in heavy print,
arcana they can profit by. I don't confess
I never cracked the spines or searched these texts.
Instead, I lie: "I rarely use them."
But they are happy, hefting the inscrutable.
My shelves are empty now,
ready to receive the great poets:
Shakespeare, Homer, Hardy, Yeats.
But my neighbors, whom I rarely see:
I think them odd, a heady sect
disposed around a pile of holy books.
Yet how must I appear to them,
enraptured by the works of Keats?
Each of us with heads bowed,
eyes fixed reverently on every word,
divided by a driveway and some trees.

One of the Spared

Suddenly a jet cleaves the air—one of those predatory fighters—and, roaring, burns a furrow down the sky. No telling where it came from—clearing the ridge at low altitude, low enough to make our hearts tremble like shook crockery. Before that, only the sound of water splashing, some small bird like a winged flute. War is never far away, or long in coming. Is that the message? What do I know of this century's horrors—one of the passed-over, one of the spared. I live in a silence only intermittently split by the sound of others' burnt cries. It's the silence into which I was born at just the right place and time. Silence like a birthright, a blessing. I should hold my tongue—such a little cup of fortune the mere passing of a jet can rattle. Far from here real jets are closing in, sniffing out a man who stands beneath them, then runs in terror. They strike him like a match against the gravel of his homeland. I think of them banking in a wide arc, unsated, adjusting their omniscient radar. But it's dangerous to think too much, dangerous to imagine—in case the silence shatters and doesn't return.

Wow & Flutter

Flaws in early hi-fi speakers:
the "WOW" of volume like a door opened
then quickly shut on a raging party;

and "flutter" was . . . well . . . flutter:
the steady *flup flup flup*
of a punctured tire.

What happened to them?
Demented cousins come to visit
in our childhood once, now gone.

They've followed their family of junk
into the grave: vacuum tubes
that glowed like small fires and gave off

an odor of burnt dust; lighted dials,
masts on a foundering ship
pitched across waves of sound.

Wow and flutter. Why do I miss them?
Now the air is a whisper of smooth notes
that come from nowhere—

each variance, each flaw
flattened out and perfect
in a place as tedious as heaven.

The American Helmet in Provence

My Belgian neighbor finds it in the woods and brings it to me as a gift, as if returning something long-borrowed and forgotten, something that belongs to me, though I fumble with it, find it hard to hold, an awkward thing I marvel at and put away, finally, on my desk.

Later, when I have time, I take a look: it's well preserved, the color of rotting leather, old wheel-drums, pitted with rust across the dome. The rim is split in two places, giving it an air of cheapness and vulnerability, as if it were a toy, though its mud-glazed shape rings with a dull peal when I tap it and it sits in my hands like some enormous shell abandoned along a beach.

When I brush my fingers across it, it makes a hissing sound only metal can make formed into a bell, a thin whisper that stops the moment you stop, an enigmatic seething, like fire. When I look inside, it's empty, just a thick wad of mud at the crown, and two square metal fixtures on the edge where the strap used to hang.

But here—and I'm not making this up—a bullet hole the size of one's finger thrust through by the left temple. And on the back, someone has scribed a Nazi cross in powdery, fading chalk. I pick it up again, and put it down, then go off to wash my hands that feel papery and light.

I sit outside to warm myself in the sun. The moon is already up—pale, translucent—like something in an X ray. I study the trunk of an olive tree, a drab sheen that spreads through bark and leaf, exhausted green the worn color of money or old fatigues. The woods are quiet, except for the cry of a single bird whose call I don't know.

White, Middle-Class, Male

I'm tired of those lethal words that hiss:
racist, capitalist, chauvinist, misogynist,
tired of constantly accusing myself—
though anyone is some of these, in part,
which simply living brings about.
Yet what if you could puncture a man
        and let the poison rush out:
masochist, sadist, narcissist, opportunist.
        Soon, he'd be diminished.
His knees would buckle, his arms shrink,
sentimentalist, moralist, romanticist, idealist.
        Purged of all those toxins,
he'd be like the witch in Oz
disappearing into a pile of laundry
—hedonist, sensualist, onanist, exhibitionist—
as if growth had been nothing but expanding hate
        and the body a blister of wickedness.

And when the hissing stopped,
he'd be a child again, pure ingot of virtue,
        model of original love.
But soon the world would coo
        pretty baby, pliant one
and he'd reach out once more to take it in.

Dear Reader, I Mean No Harm

Dear Reader, I Mean No Harm

Dear Reader, I mean no harm
but someone's out for your blood.

Maybe it's personal, the friend
of a boyfriend of an ex-lover
who's heard what a dickhead you are
and means to settle things.

Maybe it's generic—zealots
from a neighboring tribe
who happen to think your God's
the reason for their suffering.

And what of pure chance,
the certified crazy with a gun
finding you among others
stampeding his rage.

Whatever the case: beware.

Someone's lying awake as you
read this, sharpening his knife,
dreaming of the sweet butter
your throat would make.

Someone's devising a small box,
touching one thin wire
to another, making the right con-
nections to unnerve you.

Eventually
hate sits down, ready to lay

its scrupulous plans—
and your name's mentioned more than once.

This is not a theory.

Banana Knife

He called it a banana knife, and in fact
it had a long handle made of yellow pearl.
Maybe it was plastic, with wavering panes of clarity
that made it look like pearl. When he swung the blade out—
a narrow length of polished steel—
I thought a knife like that might cut through flesh
and never leave a mark, or even hurt.
So when he flicked it out to snick my arm,
I felt it brush my shirt, as though he only meant
to play a joke. I had to roll my sleeve
to watch a thread of blood outline the slit it made.
Then I noticed how the cloth was cut.
He grinned so that I'd understand
this is what he wanted: my gathering shock,
a flux of blood, however small, from the dead center of my heart.

Aliens

Early nights, the sun landed
on western ranges—nuclear, white—
then flamed red as innocent blood
behind trees utterly consumed.

We had already entered the dark
sepulchral rooms of movie-houses
as though entering our own minds
from outside, edgy as strangers.

We took our seats and sank
into long, unbreakable trances
under beams of light that dangled images
before us like a hypnotist's watch:

a typical village or small town
with its cast of petty characters
for whom we could substitute our own parents,
our own police chief and mayor.

This was the '50s, and a smothering
sameness drove us crazy, drove us
into that shivering darkness
enclosed by a greater darkness riddled with stars.

Then from somewhere among them,
from the unthinkable blackness *behind* them,
like patched-together nightmares
of all we ever feared in childhood

stepped the alien: grotesque, multi-
armed, slobbering poison, malformed
in every feature, craving only
the savory cocktail of human blood.

And we were liberated, free to behold
the face of terror, no matter how ugly,
how repulsive, unmasked now
like any vulnerable creature, one of us.

But somehow in our growing up
they changed, shed extra limbs, stopped
slobbering and learned to smile—
expressions unmistakable as love.

We must have missed their metamorphosis.
We must have been busy at the office
or the health club, busy enough
to ignore their almost Biblical conversion.

Now they glimmer on screens above us,
radiant as martyred saints;
they rise and float like angels,
like a promise letting us know the worst is over.

We're in their capable hands, we have nothing to fear.

Massive

We love to say it: *massive,* to roll the word out
like the stone rolled away from Christ's tomb,
or a mushroom of fire claiming the entire sky:
massive: it has the look and feel of something
substantial, something *ultra, deluxe,* or *super,*
hence extraordinary, hence important:
He died of a massive heart attack as if the heart
were a third-world nation with a minuscule army
overrun by a major power, its phalanxes of tanks,
its rumbling swarms of innumerable bombers.
We never say *he died of a little heart attack*
as if he had a cold, or some minor malady
that got out of hand—there's no such thing:
when the heart dies it dies entire, the way
a frown begins around the corners of a mouth
but quickly spreads until even the ears are involved,
the eyes, the jaw, the whole face
a predicate for the cool subject of disapproval.
Perhaps it's our irrepressible American spirit,
the soul of P. T. Barnum invading our chests:
heart attacks as huge as the Rockies or the heads
at Mount Rushmore, bigger than anyone else's.
We stand there awed, the sound of the word
still echoing around us like a shot that bounces
back and forth between buildings or hills,
our own fate large enough to engulf a landscape.
When have we responded so completely to a word?
And not a noble one—an abstract—but an adjective,
a mere auxilliary, like a servant, when it's the noun,
always the noun, the thing itself that kills us.

Census

There are times I have to call them together:
all the people I am, or have been,
fetching them back from discarded phases
of my life. They come like family members
to a feast, not altogether willingly,
and take their places at the table in my head:
there's Bruno, the young tough,
fond of hunting, football and barroom brawls;
and Albert, the birdwatcher, with his
hip boots and notes. There's the pale sophisticate,
Harold, in his tux, and Frank, the intellectual
with his sleeves rolled up. They come
from everywhere, like tax payers to the block:
Jules, the actor; the rock star, Lex;
Benjamin, the traveler speaking seven tongues.
The room is getting crowded, but
still they come: Gavin, the lover, Louis,
the chef, and Sidney, the author of immortal poems.
They pack together, like kids at a party.
And when I rise to address them,
they look so cooperative and meek,
so willing to give themselves up to the next
passion, the next life, they might be
converts of a new religion burning with zeal.
They don't know, even now, I betrayed them,
gave each a kiss and let them go—
though maybe it's they who betrayed me,
found wanting in the implacable purity of their eyes.

A New Age

It was the heyday of vermin:
weasel for Police Chief, mouse for Mayor.
Pestilence reigned, and frenzy.
Then came the administration of cats,
followed by the rule of dogs.
By the time of monkeys, an afternoon dolor
gripped the city. Its denizens
lay about, picking lice off their genitals.
This was remedied by the advent
of humans. Now streets were cleaned,
feces piled up and dumped
beyond the city; all hair swept clean
from neglected streets.
It was the dawning of a new age!
Then came the reign of the lowly virus.

The Committee to Upgrade Celestial Signs

meets once a year
to reevaluate old myths
that spangle heaven:
Taurus, Draco,
Perseus,
Boötes . . .
outdated in their Greek shining.
Quickly renamed,
they are reconfigured into modern shapes
—cluster by cluster—

Guitarus Major, Double Arches,
Empire State Building,
Bottle of Coke . . .
Each fall,
the firmament glitters like a new marquee,
a hit parade of celebrities
to correspond
with the season's upcoming shows:
where Cepheus glittered—
the visage of an actress
shines;
Libra morphs into the body
of a reigning hunk;
the Pleiades burn all night—
divas in a female rock group.
Trained over centuries
to forget the past
entire populations suffer
from cultural
amnesia

catalyzed by constant change—
"This is NOW!"
a favorite bumper sticker shouts,
and
"Welcome to the Interactive Cosmos!"
Constellations rise and fall,
brief as ads
that flash across the blank screen
of heaven.
Computers work around the clock
to thread stars
into relevant patterns
while last year's icons
fade
like the memory
Of someone's face
before cosmetic surgery.
At last
the Committee votes on current choices,
having sifted
through a copious Printout
of Possible Skies:
Hands go up around the table
as they nod and smile—
with the stroke of a finger
the Zodiac is realigned
against the infinite blackness behind the stars.

A '49 Merc

Someone dumped it here one night, locked
the wheel and watched it tumble into goldenrod and tansy,
ragweed grown over one door flung outward
in disgust. They did a good job, too: fenders split, windshield
veined with an intricate pattern of cracks
and fretwork. They felt, perhaps, a rare satisfaction
as the chassis crunched against rock and the rear window
buckled with its small view of the past. But the tires
are gone, and a shattered tail light shields a swarm
of hornets making home of the wreckage. How much
is enough? Years add up, placing one small burden on another
until the back yaws, shoulders slump. Whoever it was
stood here as the hood plunged over and some branches snapped,
a smell of gasoline suffusing the air, reminding us
of the exact moment of capitulation when the life
we planned can no longer be pinpointed on any map
and the way we had of getting there knocks and rattles to a halt
above a dark ravine and we go off relieved—
no, happy to be rid of the weight of all that effort and desire.

In the kingdom of old age

gods have names like Metamucil, Zantac,
        Rogaine, Viagra—redolent
        of ancient deities in the near east.

And our blood, once hot, is better thinned,
        in case a lifetime of hunger coagulates
        and stops the heart.

Each heavy sentence spells it out: “rheumatoid
        arthritis,” “myocardial infarction,”
        “dyspepsia,” “stroke”—tangled

syllables in a savage tongue our ancestors spoke.
        And buried in each cell as in
        a desert, stories of their downfall in code

        which only now we begin to crack.

How It Arrives

The sun rolls slowly from one horizon to the next
the way a cue ball touched   justso
almost doesn't make its way across a table
while shadows seep from trees like tiny leaks.
It's the way the day goes: planetary, idle.
How water, in a clear jet, doesn't seem to tumble
and the sound of locusts chirring in leaves
seems something like a pulse   but whose?
It's how the evening holds itself in long abeyance,
how light reclines on stone and gravel.
How even the temperature is late to rise
and drags its feet like someone plodding stoutly
up a thousand narrow steps at dawn.
It's how the stars arrive, like early guests at empty
houses—shy, resplendent on the lawn.

At the Health Club

This is where we go to worship—
Chapel of Toned Flesh, Church of Immaculate Bone
and Gristle—each medieval
machine an altar upon which we burn
our fat's holy tallow.
                                        Prayers we grunt
seldom rise above piped in Soul Music and Rap,
a liturgy we follow week by week in order
that we might live and not be found
wanting.
                And here, on the floor above, displays
of sushi and clear broth, a chalice of icy froth
blended from prescribed berries,
                for the body is a jealous god.

Absolved for now, and feeling good,
I step outside into snow that's been falling for hours,
collecting in gutters like cellulite, to be scraped
away later by enormous trucks,
                and every limb, anointed with the pure oil
of my effort, my whole body sheathed in sweat,
freezes.

More Things in Heaven and Earth

More Things in Heaven and Earth

I love that moment when Hamlet
turns to his friend, Horatio, and says:
There are more things in heaven and earth
Than are dreamt of in your philosophy

meaning the ghost they've just seen
and by extension the harrowing details
of the spirit-world, which no one living
may know—but he also means the secret

murder of Hamlet Senior, which accounts
for the "earth" part of Hamlet's remark—
a whole spectrum of reality about which
we are ignorant as slugs. *O day and night,*

Horatio declaims, *this is wondrous strange*
and that's the way I've always felt—
as though Hamlet wagged a finger right at me—
Thou witless oaf no wiser than a worm.

Yet it's not Prince Hamlet's quick rebuke
that rivets me, but the remark he makes
about the world: *there are more things*
in heaven and earth. The world's as various

as Horatio's response implies: stars revolve
and what we see is murder, infamy, disgrace—
enough to chill the blood and make our eyes
start out of our heads. But there's more

we cannot see, which might be beauty
or contentment, some unimagined grace
redeeming all that went before
O heaven and earth, but this is wondrous strange!

Master of Consciousness

Now you can get a "Master of Arts in Consciousness."
That's right. I saw the ad—a Master of Arts
in what you get by simply being born! Or do you?
Maybe they mean a Master of Arts
in acquired knowledge or experience—
using "consciousness" as a synonym for "knowledge,"
as in the sentence, "I'm conscious, now,
of what the Greeks accomplished in science and art."
But isn't that what education has always done,
expanding consciousness in this sense
through hard work and supervised effort?
Isn't "consciousness" what you grow into
from the moment of birth, or even conception?
Or at least from the moment the brain
becomes fully functional and begins to wonder
"Who am I, and what have I gotten myself into?"
If they mean consciousness in this sense,
they might as well be selling you your own heartbeat—
like those hucksters who sold people
what they already owned, and went off sniggering
with a suitcase full of extracted cash.
Isn't consciousness a rather limitless realm,
like an ocean, into which one dumps
all the facts and details scavanged in one's life—
an ocean one can never truly fill?
Imagine graduates of this school of "consciousness"—
when they wake up each morning,
are they more awake then you?
And if consciousness is an "art," do they do it
better than you, having been rigorously trained—
Virtuosos of Consciousness with advanced degrees
perceiving all of reality as their stage?
OK, so maybe I'm taking this too seriously
and they just mean "sensory training,"

like learning how to smell the difference
between one vintage wine and another,
or distinguishing the many subtle flavors in food.
Then they should say that clearly.
They should say: "Master of Sensory Arts."
Though it's arguable if even this—our five senses
and what we use them for—can actually
be considered an art, like Michaelangelo's *Pietà*
or Beethoven's Piano Sonata Opus 2.
And there's something about that word "Master"
that puts me off, makes me think of
black belts and gurus, eggheads with thick glasses
living their obsessessed lives beyond us all.
It's part of a new-age plot to treat us
like invalids who can't even manage awareness,
a kind of existential capitalism that makes
a commodity of our natural faculties.
Listen, I wouldn't even bother you with this
except that there are people now
who purport to teach us how to touch and taste
and feel, as if we needed technical support
for simply *being;* as if there were no such thing
as instinct or evolution, no such thing
as heredity with its iron imperatives.
It makes me crazy! Invasion of the Soul Snatchers!
Back Seat Drivers of the Psyche!
If I were you, I'd watch out: no telling how far
this thing has gone, how much of what
you do is mediated, like a puppet or a cyborg.
No telling who lies down at night with your dreams,
who rises each morning in your wife's arms.

Right & Wrong

Working as a team, Madame Kronsky and her husband,
Max, were able to isolate the elusive material
and fix it on a gel of animal fat in their laboratory at home.
After dying it with a special mixture of bee balm,
cat urine and powder made from crushed
abalone shell (an idea borrowed from the ancient Phoenicians)
they were able to study it at close range under
a microscope and witness its inscrutable behavior:
at first it turned royal purple, and pulsed
like a heart beating; but after a few hours it turned the color
of raw steak and spread itself thinner on the gel.
Within a few days, it had grown so large that a second gel
was required, and a third, until the substance
assimilated several gels an hour just to maintain its normal
rate of expansion. Soon it had taken up residence
in their living room, a dark Victorian space
Madame Kronsky decorated in the usual manner.
There, on Sundays, the Kronskys entertained visiting
colleagues—scientists of every conceivable stripe—
eager to witness the antics of this elemental jelly
that had become the center of their lives.
Now it would sit up and ooze from chair to chair,
as if following the conversation around
the room, though it showed no signs of comprehension.
Then it learned how to stretch itself into a thin
disc under the high front window that let in afternoon light.
It wasn't long before it was consuming more than
animal fat: from time to time Kronsky would toss it a cube
of sugar, or a scone, and the guests, who sat
enthralled on the edge of their chairs, exclaimed in wondering
voices on the manner in which it devoured these.
Once it flowed up into the love seat beside
Mrs. Kronsky as she was addressing the eminent psychologist,
Dr. Fritz Kriegel, from Shatzburg. "Ah," exclaimed

the doctor, “It’s molding itself to your leg! Look!”
and the other guests were able to verify Dr. Kriegel’s
excited diagnosis. Before a year had passed, the impassioned mass
had learned to draw itself up before Madame Kronsky’s
chair and lay a delicate pseudopod in her lap.
But the scientific world—the entire lay population, in fact—
was galvanized a few months later when the now
recognizable tissue climbed to the good woman’s chest, parted
the buttons on her dress, and began unmistakably
to suckle her breasts. A scandal ensued, in which authorities
attempted to wrest control of the membrane
from the Kronskys, to no avail. The capitals of Europe
were stunned, then mesmerized by this
singularly bizarre drama. Legal proceedings were initiated
by Dr. Kriegel. With a fury previously unknown
to his associates, Kriegel pressed his case with unrelenting vigor.
Possession of the thing—whatever it was: protoplasm,
ectomorph—became a focus of the courts
as attention shifted from public to private concerns.
Then one morning Max Kronsky’s heart gave out
while entertaining a group of distinguished biologists
from Moscow. Beside herself with grief—as is usual
with such intense, lifelong relationships—Madame Kronsky
soon followed. Now the press mobilized again,
as the question of what would become of the orphaned mass
piqued the curiosity of the world. This, too,
was soon resolved: it sequestered itself in the Kronsky’s
living room where it stretched lavishly
from wall to wall, covering even the heavy Victorian
furniture in its dying torpor. In a well-publicized passing
that took weeks to accomplish, the estranged tissue
died from the edges *in*, as shade by shade
a dull pallor spread itself throughout the membrane’s
now considerable bulk. Finally, a small
concentration or node of material, royal purple in hue,
was seen pulsing deep in the plasm’s midsection;
then it turned the color of uncooked meat. At once
the whole, extended substance of the thing snapped back

like a drumhead cut from its surrounding
wood. All that was left of the Kronskys' prodigy was a limp
residue of cells about the size of an ordinary
towel, colorless, undefinable and inert.
Kriegel grew melancholy, then morose.
He commissioned a special burial urn made of Lucite
so anyone could view the unfortunate
corpse. There it lay, like the skin of a full-fledged human
stripped from the body whole, until Kriegel's
own death, when it was placed in the Academy of Cautionary
Phenomena as a symbol of what can go
right in the world, and what can go wrong.

Reflections of Notre Dame de Paris

Eternité dans la Lumière

A tourist stops, frames that colossal stack of rock,
and trips the shutter: the entire building, re-configured
as light, leaps downward into a pinpoint
and flings itself across the dark to be reborn
in baptisms of fixer and soft emulsions.

Across the park another tourist filches light
from different angles: wall and buttress, peak
and spire funnel into his hands to reassemble
as a small cathedral on his desk in Minsk
or Arkansas. And all that day, from every every possible

perspective: light is garnered, processed, peeled
from bald surfaces of stone and stored away
like thought—remembered later in constellations
of precise detail: just *this* level and pitch,
this white curve, *that* depth and proportion

which align themselves in unexpected and arresting
ways, the whole venerable assemblage
with its fiery windows advancing through time.
Now multiply by weeks and years,
by generations, through the history of photography—

all those images recalled from where they've scattered,
those stone effigies illuminating rooms,
stuffed in albums or the pages of lapsed magazines;
lay them out like mirrors facing
each other across the world reflecting *ad infinitum*

only fugitive moments in the life of this church.
  But why stop there? Think of shapes
and aggregates of paint, towers of pigment that rise
  above foreshortened streets, heavy roof
lapped over windows round as palettes smeared

with hues of irridescent oil; every scrupulous
  pencil point or lump of yielding charcoal
that erects again these verticals and planes,
  these soaring arches scrawled with saints.
And every sallow block of wood carved

into architecture, any model cut or pressed
  from terra cotta, plastic, metal—
lay them beside the photographs and paintings
  before you enter the cathedral of language,
intangible, imaginary as gargoyles

that jut at shocking angles from the eaves—
  all poems, all passages in books that resurrect
these staid vaults and columns in words;
  minute, particular descriptions spun of syntax
and grammar to weave a glowing picture in your head.

Now strip those words away until you're left
  with pure image, ghostly essence
in the brain in which a monolith beside a river
  shimmers in the minds of all who ever stopped
to fix their eyes on parapet and cornice,

bell tower and relief—this infinite focus of enduring light.

Serial Position Effect

A list of things is hard to memorize,
except the first and last—
or items close enough to first and last.
The mass of items in between
gets jumbled and forgotten
as quickly as a momentary hurt or pleasure.
Try it:

tree, car, shelf, wallet, ether, wand, squabble, wrench

Now see which items you recall.
The middle fades, like a novel
that starts out with a bang
but soon falters. Or love, or hate,
or any grand infatuation we conceive as crucial.

The heart perks up at any new beginning
to memorize even the smallest polished hairs of the head,
then drops as things dissolve,
which is why so many movies
leave the hero standing at the station.

How life is like that,
from the baby's first breath
to where it's going—past the final word,
the final period
that hangs there like a doorknob
on the house of death.

In the Beginning, and the End

Anger: the world is made of it. Not molecules or atoms,
Anger. The primordial soup. The first
matrix out of which we crawled, when we learned to crawl.
Anger is the Big Bang, travels outward
shifting to red, its rubble disappearing down a black hole.
Anger gives off heat, runs down, depletes
the system. Impossible to compute Anger's location and velocity
at the same time. If you put it in a box
filled with poison, how do you know if anger is dead or alive?
Even after eons, it's possible to find
a perfect imprint of anger preserved in stone. Anger replicates,
mutates, eats its young.
Then recreates itself, straightens its spine and thinks—
I am the alpha and the omega;
nothing like me ever was
or will be . . .

Gardener

A man builds a garden on a wide hill. It's Edenesque:
fruit trees, flowers, even a stream or two. He takes his wife there
every morning to make love in the scented grass.

Soon he's bored. Why not build another garden near the first,
connected by a small path? He'll call it "Jerusalem,"
and raise a mock temple among palm trees and pools.

Now he has a place for coitus and a place for prayer.
But soon he's bored again, so he builds a third garden
and erects a forum and a circus where he slaughters animals for fun.

Now he has something to pray for! But it's not enough.
Before long, another garden rises near the last, with castles, ramparts,
even a moat. He spends his time galloping about in an iron suit.

Then he hollows out a huge lake, and has a garden built
on the opposite shore. He sails across and plants his feet on the sand:
it's a new world full of animals and dense forests.

He's happy for awhile, exploring his domain. But now
he wants to chop it all down and build a city with freeways and towers
full of sizzling light. And when he's perched in the highest tower,

he dreams of Eden and the cool trees. He thinks of making love
to his estranged wife. Tries to remember where the place was, that first
essential garden, lost among brambles and blocked paths . . .

Fisherman

A man spends his whole life fishing in himself
for something grand. It's like some lost lunker, big enough
to break all records. But he's only heard rumors, myths,
vague promises of wonder. He's only felt the shadow
of something enormous darken his life. Or has he?
Maybe it's the shadow of other fish, greater than his,
the shadow of other men's souls passing over him.
Each day he grabs his gear and makes his way
to the ocean. At least he's sure of that: or is he? Is it the ocean
or the little puddle of his tears? Is this his dinghy
or the frayed boards of his ego, scoured by storm?
He shoves off, feeling the land fall away under his boots.
Soon he's drifting under clouds, wind whispering blandishments
in his ears. It could be today: the water heaves
and settles like a chest . . . He's not far out.
It's all so pleasant, so comforting—the sunlight,
the waves. He'll go back soon, thinking: "Maybe tonight."
Night with its concealments, its shadow masking all other shadows.
Night with its privacies, its alluringly distant stars.

At the retirement home for slang

they wander around in pajamas and robes—
unkempt, malodorous, muttering to themselves—
nametags askew on soiled flannel:
Zounds! Egads! and *Shucks!*
hardly revealing the power of their former lives.
And here in a corner, sits *Balderdash!*
lost in a leather chair, his walrus mustache
breeding smoke and sputum and crumbs.
While outside, *Jimminy Crickets!* rests on the lawn
deciphering windrows of dead leaves.
But just a few miles away, near a small lake,
Camp Neologism echoes with shouts—
Far Out! and *Cool!* pick sides, as *Rad!* and *Gnarly!*
kick a hackysack around and *Awesome!*
wanders in woods discovering new fauna.
Whoa! and *Chill!* paddle a canoe,
shunning the rest, happy with their own devices.
Such gaiety is lost in summer air,
never finds its way across the lake as evening falls
shrouding boats at their moorings.
Now *Tush! Tush!* whispers to himself
as *Hogwash!* is wheeled in leaving only *Jeepers!*
singing softly in fading light. But soon
even he falls silent, until silence itself resounds
more durable than any word.

Convention: Yahoo.com

Computer eggheads cluster at the bar,
complex as circuitry,
their chatter a static I can't understand.
Name tags dangle like medals on Olympic stars.
Conversation sparkles in the foyer
flows among tables spread with food
no one seems to relish.
The future swaggers up in khaki slacks,
cell phone clapped to one ear,
laptop slung from a shoulder like an e-mailman
on the worldwide beat.
They call themselves "Netizens," and "Digeratti,"
speak in raptures of a "scalable web"
while gizmos blossom in their palms called "pilots"
and "buzz-words" sizzle in their heads.
Sexy girls in shorts and T-shirts
prance around the lobby, *Yahooligans*
blazoned on their breasts.
They only glance at me, superannuated jerk,
lone Luddite here—
Yet here I am, among the wired,
geek of language on a page,
slouching towards a night of writing in my room.

The Good Devil

He was bad at torture. Flubbed his first flaying.
Dropped his pointed trident
into a lake of oil and had to scorch himself
diving in to retrieve it. Came out looking
like a channel swimmer
sheathed in pitch. Once he stepped
on his own tail during a papal dis-
embowlment, dropped
the stomach of His Holiness
on the flagmarl where it rolled into a nearby
flue. They had to fetch it out
with iron ropes and sticks.
And once, while the other demons drew
and quartered—neatly
splitting a false prophet like a chicken—
he was busy gazing off,
admiring the tapestry of fire
that flickered on the horizon.
He missed the special Days of Profanity,
the Blasphemers' Sabbath,
the millennial Parade of Pagans.
And when that poet showed up—
the Florentine with sallow skin—he was off
gathering teeth in the Betrayer's Oven
to polish and string for his mother.
The Gossips assembled, glad for work, tongues
humming like locusts
during the first Pharonic plague.
Rumor stretched its four necks and rose on leathery wings.
When the order came up
from below, winding its way through the bowels
of authority to ordinary drudges like him,
he was banished and had to hand in
his pitchfork, tines unbloodied,

shaft still immaculate of martyrs' grease.
He had to slouch in utter shame
through the Gates of Perdition into a new
and chastening light to make his living
by the sweat of his labor—
a poor farmer now,
condemned to delve in wet earth like a simple worm.
And everything he touched throve.
Everything he planted grew
in prolific, earth-nurturing rows
glistening with everlasting life.

NOTES

"A Father's Joke" is for my sister, Gloria Edwards
"The American Helmet in Provence" is for Guy Vanparys
"Convention: Yahoo.com" is for Barry Burchell
"White, Middle-Class, Male" is for Stephen Dunn
"How It Arrives" and all the rest are for Laure-Anne Bosselaar

Kurt Brown was born in Brooklyn, New York and grew up on Long Island and in Connecticut where he attended the University of Connecticut. He spent many years in Aspen, Colorado, where he founded the Aspen Writers' Conference and edited a literary magazine, *Aspen Anthology*. His poems have appeared in *Southern Poetry Review, Massachusetts Review, Ploughshares, Harvard Review, Crazyhorse,* and many other periodicals. He is the editor of *Drive, They Said: Poems about Americans and Their Cars,* and *Verse & Universe: Poems about Science and Mathematics,* as well as a collection of essays about science and mathematics, *The Measured Word*. With his wife, poet Laure-Anne Bosselaar, he edited *Night Out: Poems about Hotels, Motels, Restaurants and Bars.* He is also the editor of three collections of lectures given at writers' conferences across America: *The True Subject, Writing It Down for James,* and *Facing the Lion.* His first full-length collection, *Return of the Prodigals,* was published by Four Way Books in 1999. He lives with his wife in New York City.